Text Message & Chat Abbreviations

CONTENTS

	Page
"A" Text Message & Chat Abbreviations	1
"B" Text Message & Chat Abbreviations	6
"C" Text Message & Chat Abbreviations	12
"D" Text Message & Chat Abbreviations	16
"E" Text Message & Chat Abbreviations	20
"F" Text Message & Chat Abbreviations	23
"G" Text Message & Chat Abbreviations	27
"H" Text Message & Chat Abbreviations	31
"I" Text Message & Chat Abbreviations	34
"J" Text Message & Chat Abbreviations	38
"K" Text Message & Chat Abbreviations	40
"L" Text Message & Chat Abbreviations	43
"M" Text Message & Chat Abbreviations	47
"N" Text Message & Chat Abbreviations	50
"O" Text Message & Chat Abbreviations	53
"P" Text Message & Chat Abbreviations	56

CONTENTS

Page

"Q" Text Message & Chat Abbreviations _____ 60

"R" Text Message & Chat Abbreviations _____ 62

"S" Text Message & Chat Abbreviations _____ 65

"T" Text Message & Chat Abbreviations _____ 70

"U" Text Message & Chat Abbreviations _____ 75

"V" Text Message & Chat Abbreviations _____ 78

"W" Text Message & Chat Abbreviations _____ 80

"X" Text Message & Chat Abbreviations _____ 84

"Y" Text Message & Chat Abbreviations _____ 86

"Z" Text Message & Chat Abbreviations _____ 89

" Numbers & Characters" Text Message & Chat Abbreviations _____ 91

"A" TEXT MESSAGE & CHAT ABBREVIATIONS

Text Message & Chat Abbreviations

A3 : Anytime, anywhere, anyplace

AA : Alcoholics Anonymous

AA : As above

AA : Ask about

AAF : As a matter of fact

AAF : As a friend

AAK : Asleep at keyboard

AAK : Alive and kicking

AAMOF : As a matter of fact

AAMOI : As a matter of interest

AAP : Always a pleasure

AAR : At any rate

AAS : Alive and smiling

AATK : Always at the keyboard

AAYF : As always, your friend

ABBR : Meaning abbreviation

ABC : Already been chewed

ABD : Already been done

ABT : About

ABT2 : Meaning "About to"

ABTA : Meaning "Good-bye" (signoff)

ABU : All bugged up

AC : Acceptable content

ACC : Anyone can come

ACD : ALT / CONTROL / DELETE

ACDNT : Accident (e-mail, Government)

ACE : Meaning "Marijuana cigarette"

ACK : Acknowledge

ACPT : Accept (e-mail, Government)

Text Message & Chat Abbreviations

ACQSTN : Acquisition (e-mail, Government)

ADAD : Another day, another dollar

ADBB : All done, bye-bye

ADD : Address

ADDY : Address

ADIH : Another day in hell

ADIP : Another day in paradise

ADMIN : Administrator

ADMINR : Administrator (Government)

ADN : Any day now

ADR : Address

AE : Area effect (online gaming)

AEAP : As early as possible

AF : April Fools

AF : As *Freak*

AF : Aggression factor (online gaming)

AFC : Away from computer

AFAIAA : As far as I am aware

AFAIC : As far as I am concerned

AFAIK : As far as I know

AFAIUI : As far as I understand it

AFAP : As far as possible

AFFA : Angels Forever, Forever Angels

AFJ : April Fool's joke

AFK : Away from keyboard

AFZ : Acronym Free Zone

AFPOE : A fresh pair of eyes

AGI : Meaning "Agility" (online gaming)

AH : At home

Text Message & Chat Abbreviations

AIAMU : And I am a money's uncle

AIGHT : Alright

AIR : As I remember

AISB : As it should be

AISB : As I said before

AISI : As I see it

AITR : Adult in the room

AKA : Also known as

ALCON : All concerned

ALOL : Actually laughing out loud

AMA : Ask me anything (Reddit)

AMAP : As much as possible

AMBW : All my best wishes

AML : All my love

AMOF : As a matter of fact

A/N : Author's note

AO : Anarchy Online (online gaming)

AOC : Available on cell

AOE : Area of effect (online game)

AOM : Age of majority

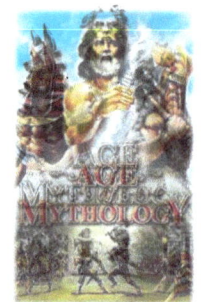

AOM : Age of Mythology (online gaming)

AOTA : All of the above

AOYP : Angel on your pillow

APAC : All praise and credit

APP : Application

APP : Appreciate

AQAP : As quick (or quiet) as possible

ARC : Archive (compressed files)

ARE : Acronym rich environment

Text Message & Chat Abbreviations

ARG : Argument

ASIG : And so it goes

ASAP : As soon as possible

A/S/L : Age/sex/location

ASL : Age/sex/location

ASLA : Age/sex/location/availability

AT : At your terminal

ATB : All the best

ATEOTD : At the end of the day

ATM : At the moment

ATSITS : All the stars in the sky

ATSL : Along the same line (or lines)

AWC : After awhile crocodile
AWESO : Awesome
AWOL : Away without leaving
AWOL : Absent without leave
AYDY : Are you done yet?
AYBABTU : All your bases belong to us (online gaming)
AYEC : At your earliest convenience
AYOR : At your own risk
AYSOS : Are you stupid or something?
AYS : Are you serious?
AYT : Are you there?
AYTMTB : And you're telling me this because
AYV : Are you vertical?

AYW : As you were

AYW : As you want / As you wish

AZN : Asian

"B" TEXT MESSAGE & CHAT ABBREVIATIONS

B : Back

B : Be

B& : Banned

B2B : Business-to-business

B2C : Business-to-consumer

B2W : Back to work

B8 : Bait (person teased or joked with, or under-aged person/teen)

B9 : Boss is watching

B/F : Boyfriend

B/G : Background (personal information request)

B4 : Before

B4N : Bye for now

BAG : Busting a gut

BAb: Bad *a*

BAE : Before anyone else

BAE : Meaning "Babe or baby"

BAFO : Best and final offer

BAK : Back at keyboard

BAM : Below average mentality

BAMF : Bad *a* mother *f*

BAO : Be aware of

BAS : Big 'butt' smile

BASIC : Meaning "Anything mainstream"

BASOR : Breathing a sigh of relief

BAU : Business as usual

BAY : Back at ya

BB : Be back

BB : Big brother
BB : Bebi / Baby (Spanish SMS)
BBC : Big bad challenge
BBIAB : Be back in a bit
BBIAF : Be back in a few
BBIAM : Be back in a minute
BBIAS : Be back in a sec

BBL : Be back later
BBN : Bye, bye now
BBQ : Barbeque, "Ownage", shooting score/frag (online gaming)
BBS : Be back soon
BBT : Be back tomorrow
BC : Because
B/C : Because

BC : Be cool
BCNU : Be seeing you
BCOS : Because
BCO : Big crush on
BCOY : Big crush on you
BD : Big deal

BDAY : Birthday
B-DAY : Birthday
BDN : Big darn number
BEG : Big evil grin
BELF : Meaning "Blood Elf" (online gaming)
BF : Boyfriend
BF : Brain fart
BFAW : Best friend at work

8

Text Message & Chat Abbreviations

BF2 : Battlefield 2 (online gaming)
BF : Best friend
BFF : Best friends forever
BFFL : Best friends for life
BFFLNMW : Best friends for life, no matter what
BFD : Big freaking deal
BFG : Big freaking grin
BFFN : Best friend for now
BFN : Bye for now
BG : Big grin
BGWM : Be gentle with me
BHL8 : Be home late
BIB : Boss is back
BIBO : Beer in, beer out
BIC : Butt in chair
BIF : Before I forget
BIH : Burn in hell
BIL : Brother in law
BIO : Meaning "I'm going to the bathroom or bathroom break"
BION : Believe it or not
BIOYA : Blow it out your *a*
BIOYN : Blow it out your nose
BIS : Best in slot (online gaming)
BISFLATM : Boy, I sure feel like a turquoise monkey!
BITMT : But in the meantime
BL : Belly laugh
BLNT : Better luck next time
Bloke : Meaning man (discord)

Text Message & Chat Abbreviations

BM : Bite me

BME : Based on my experience

BM&Y : Between me and you

BOB : Back off *buddy*

BN : Bad news

BOE : Meaning "Bind on equip" (online gaming)

BOHICA : Bend over here it comes again

BOL : Best of luck

BOM : *B* of mine

BOLO : Be on the look out

BOOMS : Bored out of my skull

BOP : Meaning "Bind on pickup" (online gaming)

BOSMKL : Bending over smacking my knee laughing

BOT : Back on topic

BOT : Be on that

BMS : Broke my scale, used when rating someone

BOYF : Boyfriend

BPLM : Big person little mind

BRB : Be right back

BR : Best regards

BRBB : Be right back *b*

BRNC : Be right back, nature calls

BRD : Bored

BRH : Be right here

BRT : Be right there

BSF : But seriously folks

BSOD : Blue screen of death

BSTS : Better safe than sorry

Text Message & Chat Abbreviations

BT : Bite this

BT : Between technologies

BTA : But then again

BTDT : Been there, done that

BTW : By the way

BTYCL : Meaning "Bootycall"

BUBU : Slang term for the most beautiful of women

BURN : Used to reference an insult

Buff : Meaning "Changed and is now stronger" (online gaming)

BWL : Bursting with laughter

BYOB : Bring your own beer

BYOC : Bring your own computer

BYOD : Bring your own device

BYOH : Bat you on (the) head

BYOP : Bring your own paint (paintball)

BYTM : Better you than me

"C" TEXT MESSAGE & CHAT ABBREVIATIONS

C&G : Chuckle & grin

C4N : Meaning "Ciao for now"

CAD : Control + Alt + Delete

CAD : Short for Canada/Canadian

Cakeday : Meaning "Birthday" (Reddit)

CAM : Camera (SMS)

CB : Coffee break

CB : Chat break

CB : Crazy *b*

CD9 : Code 9, Meaning "Parents are around"

CFS : Care for secret?

CFY : Calling for you

CHK : Check

CIAO : Good-bye (Italian word)

CICO : Coffee in, coffee out

CID : Crying in disgrace

CID : Consider it done

CLAB : Crying like a baby

CLD : Could

CLK : Click

CM : Call me

CMAP : Cover my *a* partner (online gaming)

CMB : Call me back

CMGR : Meaning "Community manager"

CMIIW : Correct me if I'm wrong

CMON : Come on

CNP : Continued (in) next post

COB : Close of business

COH : City of Heroes (online gaming)

Text Message & Chat Abbreviations

COS : Because

C/P : Cross post

CP : Chat post (or continue in IM)

CR8 : Create

Cray : Meaning "Crazy"

CRE8 : Create

CRA CRA : Slang term meaning "Crazy"

CRAFT : Can't remember a *freaking* thing

CRB : Come right back

CRBT : Crying really big tears

CRIT : Meaning "Critical hit" (online gaming)

CRZ : Crazy

CRS : Can't remember *stuff*

CSG : Chuckle, snicker, grin

CSL : Can't stop laughing

CSS : Counter-Strike Source (online gaming)

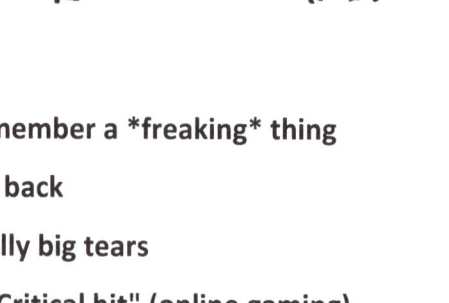

CT : Can't talk

CTC : Care to chat?

CTHU : Cracking the *heck* up

CTN : Can't talk now

CTO : Check this out

CU : See you

CU2 : See you too

CUA : See you around

CUL : See you later

CULA : See you later alligator

CUL8R : See you later

CUMID : See you in my dreams

CURLO : See you around like a donut

Text Message & Chat Abbreviations

CWD : Comment when done

CWOT : Complete waste of time

CWYL : Chat with you later

CX : Meaning "Correction"

CYA : See you

CYAL8R : See you later

CYE : Check your e-mail

CYEP : Close your eyes partner (online gaming)

CYO : See you online

"D" TEXT MESSAGE & CHAT ABBREVIATIONS

Text Message & Chat Abbreviations

D2 : Dedos / fingers (Spanish SMS)
D46? : Down for sex?
DA : Meaning "The"
DAE : Does anyone else?
DAFUQ : (What) the *freak*?
DAM : Don't annoy me
DAoC : Dark Age of Camelot (online gaming)
DBAU : Doing business as usual
DBEYR : Don't believe everything you read
DC : Disconnect
DD : Dear (or Darling) daughter
DD : Due diligence

DDG : Drop dead gorgeous
BEEZ NUTZ : A phrase used in online chat to annoy or frustrate those involved in the conversation.
DEGT : Dear (or Darling) daughter
DERP : Meaning "Stupid or silly"
DEGT : Don't even go there

DFL : Dead *freaking* last (online gaming)
DGA : Don't go anywhere
DGAF : Don't give a *freak*
DGT : Don't go there
DGTG : Don't go there, girlfriend

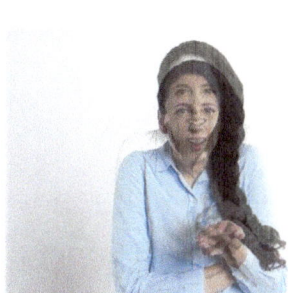

DGYF : Dang, girl you fine
DH : Dear (or Darling) husband
DHU : Dinosaur hugs (used to show support)
DIK : Darned if I know
DIKU : Do I know you?
DILLIGAF : Do I look like I give a *freak*?

Text Message & Chat Abbreviations

DILLIGAS : Do I look like I give a sugar?

DIS : Did I say?

DITYID : Did I tell you I'm distressed?

DIY : Do it yourself

DKDC : Don't know, don't care

DKP : Dragon kill points (online gaming)

D/L : Download

DL : Download

DL : Down low

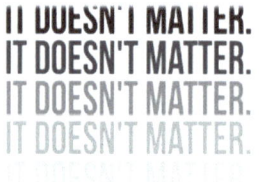

DL : Dead link

DLBBB : Don't let (the) bed bugs bite

DLTBBB : Don't let the bed bugs bite

DM : Doesn't matter

DM : Direct message (Twitter slang)

DM : Do me

DM : Dungeon Master (online gaming)

DMNO : Dude Man No Offense

DMY : Don't mess yourself

DN : Down

DNC : Do not compute (meaning "I do not understand")

DNR : Dinner (SMS)

DNT : Don't

d00d : Dude

DOE : Daughter of Eve

DORBS : Meaning "Adorable"

DOT : Damage over time (online gaming)

Downvote : Voting negatively on a thread using Reddit's voting system (Reddit)

DPS : Damage per second (online gaming)

Text Message & Chat Abbreviations

DQMOT : Don't quote me on this

DR : Didn't read

DS : Dear (or Darling) son

DTR : Define the relationship

DTRT : Do the right thing

DTS : Don't think so

DTTD : Don't touch that dial

DUPE : Duplicate

DUR : Do you remember?

DV8 : Deviate

DW : Dear (or Darling) wife

DWF : Divorced white female

DWM : Divorced white male

DXNRY : Dictionary

DYNWUTB : Do you know what you are talking about?

DYFI : Did you find it?

DYFM : Dude, you fascinate me

DYJHIW : Don't you just hate it when...?

DYOR : Do your own research (common stock market chat slang)

"E" TEXT MESSAGE & CHAT ABBREVIATIONS

Text Message & Chat Abbreviations

E : Ecstasy / Enemy (online gaming)

E1 : Everyone

E123 : Easy as one, two, three

E2EG : Ear to ear grin

EAK : Eating at keyboard

EBKAC : Error between keyboard and chair

ED : Erase display

EF4T : Effort

EG : Evil grin

EI : Eat it

EIP : Editing in progress

ELI5 : Explain like I'm 5

EM : E-mail (Twitter slang)

EMA : E-mail address (Twitter slang)

EMBAR : Meaning "Embarrassing"

EMFBI : Excuse me for butting in

EMFJI : Excuse me for jumping in

EMSG : E-mail message

ENUF : Enough

EOD : End of day

EOD : End of discussion

EOL : End of lecture

EOL : End of life

EOM : End of message

EOS : End of show

EOT : End of transmission

EQ : EverQuest (online gaming)

ERP : Meaning "Erotic Role-Play" (online gaming)

ERS2 : Eres tz / are you (Spanish SMS)

ES : Erase screen

ESAD : Eat *S* and die!

ETA : Estimated time (of) arrival

ETA : Edited to add

EVA : Ever

EVO : Evolution

EWG : Evil wicked grin (in fun, teasing)

EWI : Emailing while intoxicated

EXTRA : Meaning over the top

EYC : Excitable, yet calm

EZ : Easy

EZY : Easy

"F" TEXT MESSAGE & CHAT ABBREVIATIONS

F : Meaning "Female"

F2F : Face to face

F2P : Free to play (online gaming)

F4F : Follow for follow (Instagram)

FAAK : Falling asleep at keyboard

FAB : Fabulous

Facepalm : Used to represent the gesture of "smacking your forehead with your palm" to express embarrassment or frustration

FAF : Funny as *freak*

FAM : Family, but not limited to actual family members. Could mean friends.

FAQ : Frequently asked questions

FAY : *Freak* all you

FB : Facebook

FBB : Meaning "Facebook buddy"

FBC : Facebook chat

FBF : Flashback Friday

FBF : Meaning "Facebook friend"

FBF : Fat boy food (e.g. pizza, burgers, fries)

FBFR : Facebook friend

FBM : Fine by me

FBO : Facebook official (An official update from Facebook)

FBOW : For better or worse

FC : Fingers crossed

FC : Full card (online gaming)

FCINGO : For crying out loud

FCOL : For crying out loud

Feelsbadman : A social meme that means to feel negative.

Feelsbatman : A social meme taking "feelsbadman" to the extreme. This references the DC super hero Batman because he witnessed the murder of his parents.

Feelsgoodman : A social meme that means to feel positive.

FEITCTAJ : *Freak* 'em if they can't take a joke

FF : Follow Friday (Twitter slang)

FFA : Free for all (online gaming)

FFS : For *freak's* sake

FICCL : Frankly I couldn't care any less

FIF : *Freak* I'm funny

FIIK : *Freaked* if I know

FIIOOH : Forget it, I'm out of here

FIL : Father in law

FIMH : Forever in my heart

Finna : Meaning "Going to"

Finsta : A second Instagram account where someone can post things that they're too afraid to post on their main account.

FISH : First in, still here

FITB : Fill in the blank

FML : *Freak* My Life

FOMC : Falling off my chair

FOMO : Fear of missing out (definition)

FOAD : *Freak* off and die

FOAF : Friend of a friend

FOMCL : Falling off my chair laughing

FRT : For real though

FTBOMH : From the bottom of my heart

FTFY : Fixed that for you

FTL : For the loss

FTW : For the win

FU : *Freak* you

FUBAR : Fouled up beyond all recognition

FUBB : Fouled up beyond belief

FUD : Face up deal (online gaming)

FUTAB : Feet up, take a break

FW : Forward

FWB : Friend with benefits

FWIW : For what it's worth

FWM : Fine with me

FWP : First world problems

FYE : Fire, something that is cool

FYEO : For your eyes only

FYA : For your amusement

FYA : For your action

FYI : For your information

"G" TEXT MESSAGE & CHAT ABBREVIATIONS

Text Message & Chat Abbreviations

G : Grin

G : Giggle

G+ : Google+

G/F : Girlfriend

G2CU : Good to see you

G2G : Got to go

G2GICYAL8ER : Got to go I'll see you later

G2R : Got to run

G2TU : Got to tell u (you)

G4C : Going for coffee

G9 : Genius

GA : Go ahead

GAC : Get a clue

GAFC : Get a *freaking* clue

GAL : Get a life

GANK : Meaning "A player ambush or unfair player kill" (online gaming)

GAS : Got a second?

GAS : Greetings and salutations

GB : Goodbye

GBTW : Get back to work

GBU : God bless you

GD : Good

GDR : Grinning, ducking, and running

GD/R : Grinning, ducking, and running

GFI : Go for it

GF : Girl friend

GFN : Gone for now

GG : Gotta Go

GG : Good Game (online gaming)

Text Message & Chat Abbreviations

GG : Brother (Mandarin Chinese)

GGA : Good game, all (online gaming)

GGE1 : Good game, everyone (online gaming)

GGU2 : Good game, you too

GGMSOT : Gotta get me some of that

GGOH : Gotta Get Outta Here

GGP : Got to go pee

GH : Good hand (online gaming)

GIAR : Give it a rest

GIC : Gift in crib (online gaming)

GIGO : Garbage in, garbage out

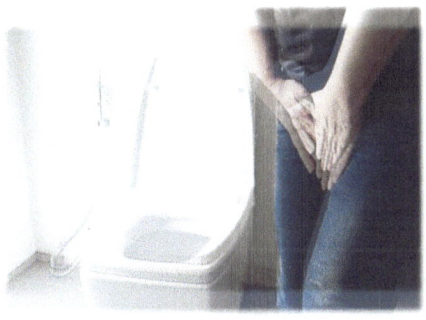

GIRL : Guy in real life

GJ : Good job

GL : Good luck

GL2U : Good luck to you (online gaming)

GLA : Good luck all (online gaming)

GL/HF : Good luck, have fun (online gaming)

GLE : Good luck everyone (online gaming)

GLE1 : Good luck everyone (online gaming)

GLNG : Good luck next game (online gaming)

GMBA : Giggling my butt off

GMTA : Great minds think alike

GMV : Got my vote

GN : Good night

GNA : Good night all

GNE1 : Good night everyone

GNIGHT : Good night

GNITE : Good night

GNSD : Good night, sweet dreams

GOAT : Greatest of all time(s)

GOI : Get over it

GOL : Giggling out loud

GOMB : Get off my back

GPOY : Gratuitous picture of yourself

GR8 : Great

GRATZ : Congratulations

GRL : Girl

GRWG : Get right with God

GR&D : Grinning, running and ducking

GS : Good shot

GS : Good split (online gaming)

GT : Good try

GTFO : Get the *freak* out

GTFOH : Get the *freak* outta here

GTG : Got to go

GTM : Giggling to myself

GTRM : Going to read mail

GTSY : Great (or good) to see you

GUD : Good

GWHTLC : Glad we had this little chat

"H" TEXT MESSAGE & CHAT ABBREVIATIONS

H : Hug

H8 : Hate

H8TTU : Hate to be you

HAG1 : Have a good one

HAK : Hug and kiss

HALP : Help (Discord)

HAU : How about you?

H&K : Hugs & kisses

H2CUS : Hope to see you soon

HAGN : Have a good night

HAGO : Have a good one

HAND : Have a nice day

HAWD : Have a wonderful day

HAWT : Meaning "Sexy" or "Attractive"

HB : Hurry back

HB : Hug back

HBB : Happy birthday

H-BDAY : Happy Birthday

HBU : How about you?

HF : Have fun

HFAC : Holy flipping animal crackers

H-FDAY : Happy Father's Day

HHIS : Head hanging in shame

HIFW : How I felt when... (Used with photo or gif animation)

HL : Half Life (online gaming)

HLA : Hola / hello (Spanish SMS)

H-MDAY : Happy Mother's Day

HMU : Hit me up

HNL : (w)Hole (a)Nother Level

Text Message & Chat Abbreviations

HOAS : Hold on a second

HP : Hit points / Health points (online gaming)

HRU : How are you?

HTH : Hope this helps

HUB : Head up butt

HUYA : Head up your *butt*

HV : Have

HVH : Heroic Violet Hold (online gaming)

HW : Homework

HYFR : Hell yeah, *freaking* right!

"I" TEXT MESSAGE & CHAT ABBREVIATIONS

Text Message & Chat Abbreviations

I2 : I too (me too)

IA8 : I already ate

IAAA : I am an accountant

IAAD : I am a doctor

IAAL : I am a lawyer

IAC : In any case

IAE : In any event

IANAC : I am not a crook

IANAL : I am not a lawyer

IAO : I am out (of here)

IB : I'm back

IC : I see

ICAM : I couldn't agree more

ICBW : It could be worse

ICEDI : I can't even discuss it

ICFILWU : I could fall in love with you

ICYMI : In case you missed it (Twitter slang)

IDBI : I don't believe it

IDC : I don't care

IDGAF : I don't give a *freak*

IDK : I don't know

IDTS : I don't think so

IDUNNO : I don't know

IFYP : I feel your pain

IG : Instagram

IG2R : I got to run

IGHT : I got high tonight

IGN : I (I've) got nothing

IGP : I got to (go) pee

IHNI : I have no idea
IIRC : If I remember correctly
IIIO : Intel inside, idiot outside
IK : I know
IKR : I know, right?
ILBL8 : I'll be late
ILU : I love you
ILUM : I love you man
ILYSM : I love you so much
ILY : I love you
IM : Instant message
IMAO : In my arrogant opinion
IMHO : In my humble opinion
ImL : (in Arial font) Means I love you (a way of using the American sign language in text)
IMNSHO : In my not so humble opinion
IMO : In my opinion
IMS : I am sorry
IMSB : I am so bored
IMTM : I am the man
IMU : I miss u (you)
INAL : I'm not a lawyer
INC : Meaning "Incoming" (online gaming)
INV : Meaning "Invite"
IOMH : In over my head
IOW : In other words
IRL : In real life
IRMC : I rest my case
ISLY : I still love you
ISO : In search of

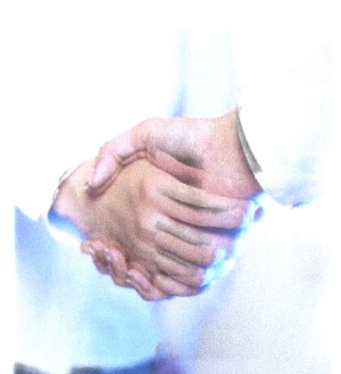

ITAM : It's The Accounting, Man (financial blogs)
ITT : In This Thread
ITYK : I thought you knew
IUSS : If you say so
IWALU : I will always love you
IWAWO : I want a way out
IWIAM : Idiot wrapped in a moron
IWSN : I want sex now
IYKWIM : If you know what I mean
IYO : In your opinion
IYQ : Meaning "I like you"
IYSS : If you say so

"J" TEXT MESSAGE & CHAT ABBREVIATIONS

Text Message & Chat Abbreviations

j00 : You

j00r : Your

JAC : Just a second

JAM : Just a minute

JAS : Just a second

JC (J/C) : Just checking

JDI : Just do it

JELLY : Meaning "Jealous"

JFF : Just for fun

JFGI : Just *freaking* Google it

JIC : Just in case

JJ (J/J) : Just joking

JJA : Just joking around

JK (J/K) : Just kidding

JLMK : Just let me know

JMO : Just my opinion

JP : Just playing

JP : Jackpot (online gaming, bingo games)

JT (J/T) : Just teasing

JTLYK : Just to let you know

JV : Joint venture

JW : Just wondering

39

"K" TEXT MESSAGE & CHAT ABBREVIATIONS

Text Message & Chat Abbreviations

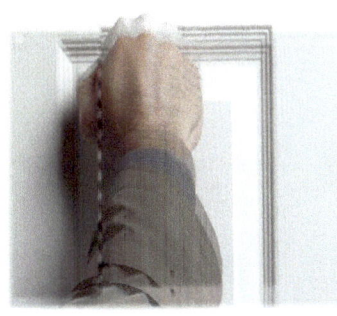

K : Okay

KK : Knock, knock

KK : Okay, Okay!

k/b : Keyboard

KB : Keyboard

KB : Kick butt (online gaming)

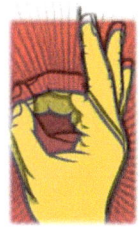

KBFU : Means Cracking (K) the (B as in Da) *freak* up

 KEWL : Cool

KEYA : I will key you later

KEYME : Key me when you get in

KFY : Kiss for you

KIA : Know it all

KIR : Keep it real

KISS : Keep it simple, stupid

KIT : Keep in touch

KMA : Kiss my *a*

KMK : Kiss my keister

KMS : Killing myself

KMT : Kiss my tushie

KOC : Kiss on cheek

KOL : Key opinion leader

Koreaboo : Someone obsessed with Korean culture

KOS : Kid over shoulder

KOS : Kill on sight (online gaming)

KOW : Knock on wood

KOTC : Kiss on the cheek

KOTD : Kicks of the day (Instagram)

KOTL : Kiss on the lips

41

KNIM : Know what I mean?

KNOW : Meaning "Knowledge"

KPC : Keeping parents clueless

KS : Kill then steal (online gaming)

KSC : Kind (of) sort (of) chuckle

KUTGW : Keep up the good work

KYS : Kill yourself

"L" TEXT MESSAGE & CHAT ABBREVIATIONS

L2G : Like to go?
L2G : (would) Love to go
L2K : Like to come
L2P : Learn to play
l33t : Leet, meaning "Elite"
L4L : Like for like (Instagram)
L8R : Later
L8RG8R : Later, gator

LAB : Life's a bitc*
LBAY : Laughing back at you
LBS : Laughing, but serious
LBVS : Laughing, but very serious
LD : Later, dude
LD : Long distance
LDO : Like, duh obviously
LEMENO : Let me know

LERK : Leaving easy reach of keyboard
LFD : Left for day
LFG : Looking for group (online gaming)
LFG : Looking for guard (online gaming)
LFM : Looking for more (online gaming)
LGH : Let's get high
LH6 : Let's have s*x
LHSX : Let's have s*x
LHM : Lord help me
LHO : Laughing head off
LI : LinkedIn
LIC : Like I care
LIK : Meaning "Liquor"

44

Text Message & Chat Abbreviations

LIMT : Laugh in my tummy

LIT : Meaning really good or something fun and exciting

LIT : Extremely intoxicated

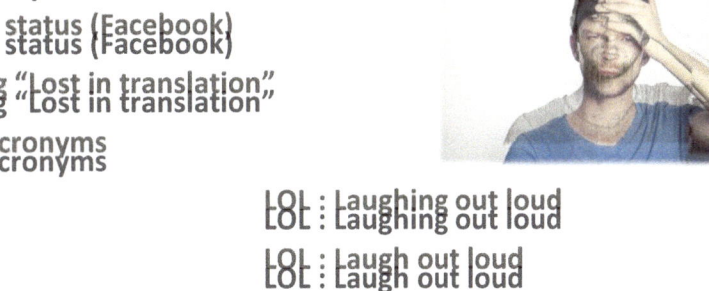

LLGB : Love, later, God bless

LLS : Laughing like *silly*

LMAO : Laughing my *a* off

LMBO : Laughing my butt off

LMFAO : Laughing my freaking *a* off

LMIRL : Let's meet in real life

LMK : Let me know

LMMFAO : Laughing my mother freaking a** off

LMNK : Leave my name out

LMS : Like my status (Facebook)

LNT : Meaning "Lost in translation"

LOA : List of acronyms

LOL : Laughing out loud

LOL : Laugh out loud

LOL : Lots of love

LOLH : Laughing out loud hysterically

LOLO : Lots of love

LOLWTF : Laughing out loud (saying) "What the *freak*?"

LOTI : Laughing on the inside

LOTR : Lord of The Rings (online gaming)

LQTM : Laughing quietly to myself

LSHMBH : Laugh so hard my belly hurts

LSV : Language, sex and violence

LTD : Living the dream

LTLWDLS : Let's twist like we did last summer

45

LTNS : Long time no see

LTOD : Laptop of death

LTS : Laughing to self

LULT : Love you long time

LULZ : Laughing (joke)

LVM : Left voice mail

LWOS : Laughing without smiling

LY : Love you or ya

LYLAS : Love you like a sis

LYLC : Love you like crazy

LYSM : Love you so much

"M" TEXT MESSAGE & CHAT ABBREVIATIONS

M$: Microsoft

M8 : Mate

MB : Mamma's boy

MBS : Mom behind shoulder

MC : Merry Christmas

MDIAC : My Dad is a cop

 MEGO : My eyes glaze over

MEH : Meaning "A shrug or shrugging shoulders"

MEH : Meaning "A so-so or just okay"

MEHH : Meaning "A sigh or sighing"

MEZ : Meaning "Mesmerize" (online gaming)

MFI : Mad for it

MFW : My face when... (Used with photo or gif)

MGB : May God bless

MGMT : Management

MHOTY : My hat (is) off to you

MIRL : Me in real life

MIRL : Meet in real life

MISS.(number) : Meaning "Child and her age" (Miss.3 would be a 3-year old daughter)

MKAY : Meaning "Mmm, okay"

MLM : Meaning "Give the middle finger"

MM : Sister (Mandarin Chinese)

MMK : Meaning "Okay?" (as a question)

MNC : Mother nature calls

MNSG : Mensaje (message in Spanish)

MOD : Moderator

MOD : Modification (online gaming)

MORF : Male or female?

MOMBOY : Mamma's boy

MOO : My own opinion

MOOS : Member of the opposite sex

MOS : Mother over shoulder

MOSS : Member of same sex

MP : Mana points (online gaming)

MR.(number) : Meaning "Child and his age" (Mr.3 would be a 3-year old son)

MRT : Modified ReTweet (Twitter slang)

MRW : My reaction when... (Used with photo or gif)

MSG : Message

MTF : More to follow

MTFBWU : May the force be with you

MU : Miss U (you)

MUAH : Multiple unsuccessful attempts (at/to) humor

MUSM : Miss you so much

MWAH : Meaning "Kiss" (it is the sound made when kissing through the air)

MYO : Mind your own (business)

MYOB : Mind your own business

"N" TEXT MESSAGE & CHAT ABBREVIATIONS

n00b : Newbie

N1 : Nice one

N2M : Nothing too much

NADT : Not a darn thing

NALOPKT : Not a lot of people know that

NANA : Not now, no need

NBD : No big deal

NBFAB : No bad for a beginner (online gaming)

NC : Nice crib (online gaming)

ND : Nice double (online gaming)

NE : Any

NE1 : Anyone

NERF : Meaning "Changed and is now weaker" (online gaming)

NFM : None for me / Not for me

NGL : Not gonna (going to) lie

NFS : Need for Speed (online gaming)

NFS : Not for sale

NFW : No *freaking* way

NFW : Not for work

NFWS : Not for work safe

NH : Nice hand (online gaming)

NIFOC : Naked in front of computer

NIGI : Now I get it

NIMBY : Not in my back yard

NIROK : Not in reach of keyboard

NLT : No later than

NM : Nothing much

NM : Never mind

NM : Nice meld (online gaming)

NMH : Not much here
NMJC : Nothing much, just chilling
NMU : Not much, you?
NO1 : No one
NOOB : Meaning "Someone who is bad at (online) games"

NOWL : Meaning "Knowledge"
NOYB : None of your business
NP : No problem
NPC : Non-playing character (online gaming)
NQT : Newly qualified teacher
NR : Nice roll (online gaming)
NRN : No response/reply necessary
NS : Nice score (online gaming)
NS : Nice split (online gaming)
NSA : No strings attached
NSFL : Not safe for life
NSFW : Not safe for work
NSISR : Not sure if spelled right
NT : Nice try
NTHING : Nothing (SMS)
NTS : Note to self
NVM : Never mind
NVR : Never
NW : No way
NWO : No way out

"O" TEXT MESSAGE & CHAT ABBREVIATIONS

Text Message & Chat Abbreviations

O4U : Only for you

O : Opponent (online gaming)

O : Meaning "Hugs"

O : Over

OA : Online auctions (see more auction abbreviations)

OATUS : On a totally unrelated subject

OB : Oh baby

OB : Oh brother

OBV : Obviously

OFC : Of course

OG : Original gangster

OGIM : Oh God, it's Monday

OH : Overheard

OHHEMMGEE : Meaning "Oh My God" (alternate of OMG)

OI : Operator indisposed

OIB : Oh, I'm back

OIC : Oh, I see

OJ : Only joking

OL : Old lady

OLL : Online love

OM : Old man

OM : Oh, my

OMAA : Oh, my aching *A* (butt)

OMDB : Over my dead body

OMFG : Oh my *freaking* God

OMG : Oh my God

OMG : Oh my gosh

OMGYG2BK : Oh my God, you got to be kidding

OMGYS : Oh my gosh, you suck

54

OMS : On my soul (to promise)

OMW : On my way

ONL : Online

OO : Over and out

OOC : Out of character

OOH : Out of here

OOMF : One of my followers

OOTD : One of these days

OOTD : Outfit of the day (Instagram)

OOTO : Out of the office

OP : On phone

ORLY : Oh really?

OS : Operating system

OT : Off topic (discussion forums)

OTB : Off to bed

OTFL : On the floor laughing

OTL : Out to lunch

OTOH : On the other hand

OTP : On the phone

OTP : One true pairing; two people you would love to see as a couple

OTT : Over the top

OTTOMH : Off the top of my head

OTW : Off to work / On the way

OVA : Over

OYO : On your own

"P" TEXT MESSAGE & CHAT ABBREVIATIONS

Text Message & Chat Abbreviations

P : Partner (online gaming)

P2P : Parent to parent

P2P : Peer to peer

P2P : Pay to play (online gaming)

P911 : Parents coming into room alert

PAP : Post a picture

PAT : Meaning "Patrol" (online gaming)

PAW : Parents are watching

PBOOK : Phonebook (e-mail)

PC : Player character (online gaming)

PCM : Please call me

PDA : Personal display (of) affection

PDH : Pretty darn happy

PDS : Please don't shoot

PDQ : Pretty darn quick

PEEPS : People

PFT : Pretty *freaking* tight

PIC : Picture

PIP : Peeing in pants (laughing hard)

PIR : Parents in room

PISS : Put in some sugar

PITA : Pain in the *butt*

PKMN : Pokemon (online gaming)

PL8 : Plate

PLD : Played

PLMK : Please let me know

PLS : Please

PLU : People like us

PLZ : Please

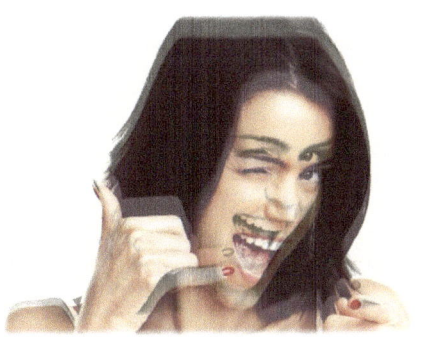

PLZTLME : Please tell me

PM : Private Message

PMFI : Pardon me for interrupting

PMFJI : Pardon me for jumping in

PMSL : Pee myself laughing

POAHF : Put on a happy face

POIDH : Picture, or it didn't happen

POS : Parent over shoulder

POT : Meaning "Potion" (online gaming)

POTD : Photo of the day (Instagram)

POV : Point of view

POV : Privately owned vehicle (originally used to differentiate an individual's vehicle from a military vehicle)

PPL : People

PPU : Pending pick-up

PRESH : Precious

PROBS : Probably

PROLLY : Probably

PROGGY : Meaning "Computer program"

PRON : Meaning "Pornography"

PRT : Party

PRT : Please Retweet (Twitter slang)

PRW : People/parents are watching

PSA : Public Service Announcement

PSOS : Parent standing over shoulder

PSP : Playstation Portable

Text Message & Chat Abbreviations

PST : Please send & tell (online gaming)

PTFO : Pass the *freak* out

PTIYPASI : Put that in your pipe and smoke it

PTL : Praise the Lord

PTMM : Please tell me more

PTO : Paid time off

PTO : Personal time off

PTO : Parent Teacher Organization

PUG : Pick up group (online gaming)

PVE : Player vs enemy, Player versus environment (online gaming)

PVP : Player versus player (online gaming)

PWN : Meaning "Own"

PXT : Please explain that

PU : That stinks!

PUKS : Pick up kids (SMS)

PYT : Pretty young thing

PZ : Peace

PZA : Pizza

"Q" TEXT MESSAGE & CHAT ABBREVIATIONS

Q : Queue
Q4U : (I have a) question for you
QC : Quality control
QFE : Question for everyone

QFI : Quoted for idiocy
QFI : Quoted for irony
QFT : Quoted for truth
QIK : Quick
QL : Quit laughing

QOTD : Quote of the day
QQ (qq) (Q_Q) : Meaning "Crying eyes"
QQ : Quick question
QSL : Reply
QSO : Conversation
QT : Cutie
QTPI : Cutie pie

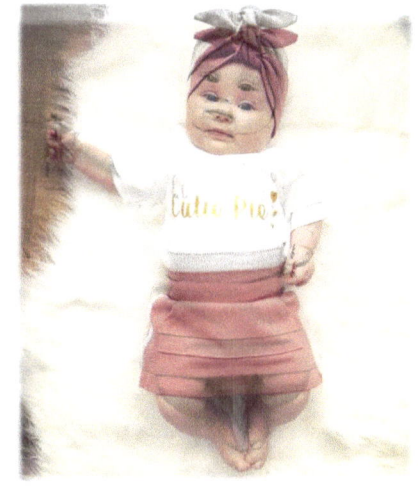

"R" TEXT MESSAGE & CHAT ABBREVIATIONS

Text Message & Chat Abbreviations

R : Meaning "Are"

R8 : Rate (SMS)

RBAY : Right back at you

RFN : Right *freaking* now

RGR : Roger (I agree, I understand)

RHIP : Rank has its privileges

RIP : Rest in peace

RL : Real life

RLY : Really

RME : Rolling my eyes

RMLB : Read my lips baby

RMMM : Read my mail man

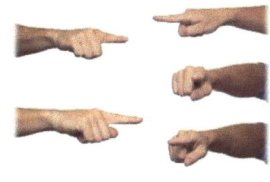

ROFL : Rolling on floor laughing

ROFLCOPTER : Rolling on floor laughing and spinning around

ROFLMAO : Rolling on the floor, laughing my *butt* off

ROTFL : Rolling on the floor laughing

ROTFLUTS : Rolling on the floor laughing unable to speak

RS : Runescape (online gaming)

RSN : Real soon now

RT : Roger that

RT : Retweet (Twitter slang)

RTBS : Reason to be single

RTFM : Read the *freaking* manual

RTFQ : Read the *freaking* question

RTHX : Meaning "Thanks for the RT (Retweet)" (Twitter slang)

RTMS : Read the manual, stupid

RTNTN : Retention (email, Government)

RTRCTV : Retroactive (email, Government)

RTRMT : Retirement (email, Government)

63

RTSM : Read the stupid manual

RTWFQ : Read the whole *freaking* question

RU : Are you?

RUMOF : Are you male or female?

RUT : Are u (you) there?

RUOK : Are you okay?

RX : Regards

RW : Real world

RX : Meaning "Drugs or prescriptions"

RYB : Read your Bible

RYO : Roll your own

RYS : Read your screen

RYS : Are you single?

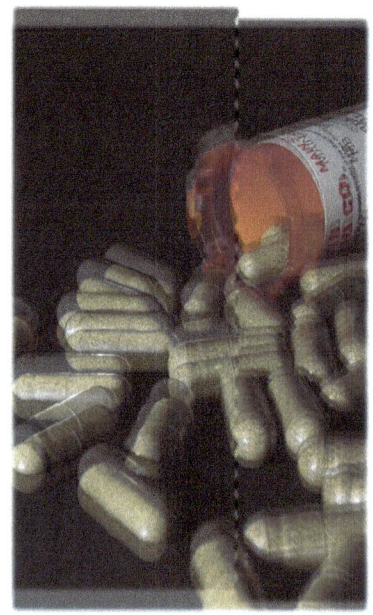

"S" TEXT MESSAGE & CHAT ABBREVIATIONS

S2R : Send to receive (meaning "Send me your picture to get mine")

S2S : Sorry to say

S4L : Spam for life

SAL : Such a laugh

SAT : Sorry about that

SAVAGE : Slang for a shockingly careless expression or response to an event or message

SB : Snap back (Snapchat)

SB : Should be

SB : Smiling back

SBIA : Meaning "Standing back in amazement"

SBT : Sorry 'bout that

SC : Stay cool

SD : Sweet dreams

SDMB : Sweet dreams, my baby

SENPAI : Meaning "Someone older than you, someone you look up to"

SEO : Search engine optimization

SETE : Smiling Ear-to-Ear

SELFIE : A photo that is taken of oneself for social media sharing (definition)

SFAIK : So far as I know

SH : Same here

SH^ : Shut up

SHID : Slapping head in disgust

SHIP : Slang for "wishing two people were in a relationship"

SICNR : Sorry, I could not resist

SIG2R : Sorry, I got to run

SIHTH : Stupidity is hard to take

SIMYC : Sorry I missed your call

SIR : Strike it rich

SIS : Snickering in silence

SIS : Meaning "Sister"

SIT : Stay in touch

SITD : Still in the dark

SJW : Social justice warrior

SK8 : Skate

SK8NG : Skating

SK8R : Skater

SK8RBOI : Skater Boy

SLAP : Sounds like a plan

Slay : To succeed at something

SM : Social media

SMAZED : Smoky haze (marijuana stoned)

SMEXI : Combination of sexy and Mexican, used to describe attractive people

SMH : Shaking my head

SMHID : Scratching my head in disbelief

SNAFU : Situation normal all fouled up

SNERT : Snot nosed egotistical rude teenager

SNR : Streaks and Recents (Snapchat)

SO : Significant other

SOAB : Son of a *B*

S'OK : Meaning "It'(s) okay (ok)"

SOL : Sooner or later

SOMY : Sick of me yet?

SorG : Straight or Gay?

SOS : Meaning "Help"

SOS : Son of Sam

SOT : Short of time

SOTMG : Short of time, must go

SOWM : Someone with me

SPK : Speak (SMS)

SRSLY : Seriously

SPST : Same place, same time

SPTO : Spoke to

SQ : Square

SRY : Sorry

SS : So sorry

SSDD : Same stuff, different day

SSIF : So stupid it's funny

SSINF : So stupid it's not funny

ST&D : Stop texting and drive

Stan : Meaning "A die-hard fan of someone" (Snapchat)

STFU : Shut the *freak* up

STR8 : Straight

STW : Search the Web

SU : Shut up

SUITM : See you in the morning

SUL : See you later

SUP : What's up?

SUTH : So use(d) to haters (Facebook)

SUX : Meanings "Sucks or it sucks"

SUYF : Shut up you fool

SWAG : Meaning "Free stuff and giveaways from tech tradeshows" (definition)

SWAG : Scientific wild *a* guess

SWAK : Sent (or sealed) with a kiss

SWALK : Sealed (or sealed) with a loving kiss

Text Message & Chat Abbreviations

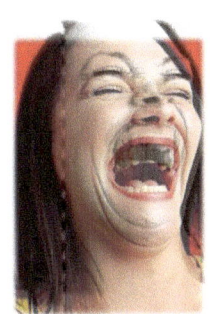

SWAT : Scientific wild *butt* guess

SWL : Screaming with laughter

SWMBO : She who must be obeyed (meaning "Wife or partner")

SYL : See you later

SYS : See you soon

SYY : Shut your yapper

"T" TEXT MESSAGE & CHAT ABBREVIATIONS

T+ : Think positive

T4BU : Thanks for being you

T:)T : Think happy thoughts

TA : Thanks a lot

TAFN : That's all for now

TAM : Tomorrow a.m.

TANK : Meaning "Really strong"

TANKED : Meaning "Owned"

TANKING : Meaning "Owning"

TARFU : Things Are Really *Fouled* Up.

TAU : Thinking about u (you)

TAUMUALU : Thinking about you miss you always love you

TBAG : Process of disgracing a corpse, taunting a fragged/killed player (online gaming)

TBBH : To be brutally honest

TBC : To be continued

TBD : To be determined

TBH : To be honest

TBL : Text back later

TBT : Throwback Thursday (Twitter slang)

TC : Take care

TCB : Take care of business

TCOY : Take care of yourself

TD : Tower defense (online gaming)

TD2M : Talk dirty to me

TDTM : Talk dirty to me

TEA : Meaning "Gossip"

TFF : Too *freaking* funny

TFS : Thanks for sharing

71

TFTF : Thanks for the follow (Twitter slang)
TFTI : Thanks for the invitation
TFTT : Thanks for this tweet (Twitter slang)
TG : Thank goodness
TGIF : Thank God it's Friday
THNQ : Thank-you (SMS)
THNX : Thanks
THOT : That wh*re over there
THT : Think happy thoughts
THX : Thanks
TIA : Thanks in advance
TIAD : Tomorrow is another day
TIC : Tongue-in-cheek
TIL : Today I learned
TILIS : Tell it like it is

TIR : Teacher in room
TLK2UL8R : Talk to you later
TL : Too long
TL;DR : Too long; didn't read
TM : Trust me

TMA : Take my advice
TMB : Text me back
TMB : Tweet me back (Twitter slang)
TMI : Too much information
TMOT : Trust me on this
TMTH : Too much to handle
TMYL : Tell me your location
TMWFI : Take my word for it
TNSTAAFL : There's no such thing as a free lunch

Text Message & Chat Abbreviations

TNT : un(T)il (N)ext (T)ime

TOJ : Tears of joy

TOS : Terms of service

TOTES : Totally

TOU : Thinking of you

TOY : Thinking of you

TPM : Tomorrow p.m.

TPTB : The powers that be

TQ : Te quiero / I love you (Spanish SMS)

TSH : Tripping so hard

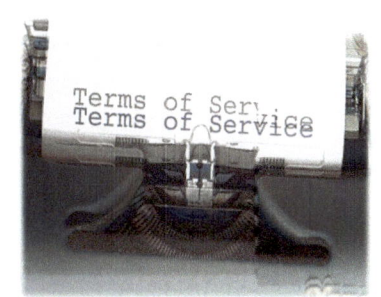

TSNF : That's so not fair

TSTB : The sooner, the better

TT : Trending topic (Twitter slang)

TTFN : Ta ta for now

TTLY : Totally

TTTT : These things take time

TTUL : Talk to you later

TU : Thank you

TUI : Turning you in

TURNT : Meaning "Excitement, turned up"

TWSS : That's what she said

TTG : Time to go

TTYAFN : Talk to you awhile from now

TTYL : Talk to you later

TTYS : Talk to you soon

TY : Thank you

TYFC : Thank you for charity (online gaming)

TYFYC : Thank you for your comment

Text Message & Chat Abbreviations

TYS : Told you so

TYT : Take your time

TYSO : Thank you so much

TYAFY : Thank you and *freak* you

TYVM : Thank you very much

TX : Thanks

"U" TEXT MESSAGE & CHAT ABBREVIATIONS

URS : Up yours

UCMU : You crack me up

UDI : Unidentified drinking injury (bruise and so on)

UDM : U (You) da (the) man

UDS : Ugly domestic scene

UFB : Un *freaking* believable

UFN : Until further notice

UFWM : You *freaking* with me?

UGTBK : You've got to be kidding

UHGTBSM : You have got to be s#$t*ing me!

UKTR : You know that's right

UL : Upload

U=L : Meaning "You will"

UNA : Use no acronyms

UN4TUN8 : Unfortunate

UNBLEFBLE : Unbelievable

UNCRTN : Uncertain

UNPC : Un- (not) politically correct

UOK : (Are) You ok?

UR : You are / You're

UR2YS4ME : You are too wise for me

URA* : You are a star

URH : You are hot (U R Hot)

URSKTM : You are so kind to me

URTM : You are the man

URW : You are welcome

USBCA : Until something better comes along

USU : Usually

UT : Unreal Tournament (online gaming)

Text Message & Chat Abbreviations

UT2L : You take too long
UTM : You tell me
UV : Unpleasant visual
UW : You're welcome
UX : User experience

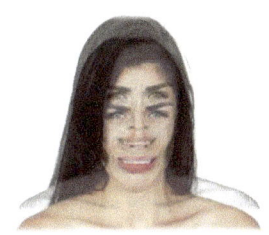

"V" TEXT MESSAGE & CHAT ABBREVIATIONS

Text Message & Chat Abbreviations

V : Very

VAT : Value added tax

VBL : Visible bra line

VPL : Visible panty line

VBS : Very big smile

VC : Voice chat

VEG : Very evil grin

VFF : Very freaking funny

VFM : Value for money

VGC : Very good condition

VGG : Very good game (online gaming)

VGH : Very good hand (online gaming)

VIP : Very important person

VM : Voice mail

VN : Very nice

VNH : Very nice hand (online gaming)

VoIP : Voice over Internet Protocol (definition)

VRY : Very

VSC : Very soft chuckle

VSF : Very sad face

VWD : Very well done (online gaming)

VWP : Very well played (online gaming)

"W" TEXT MESSAGE & CHAT ABBREVIATIONS

W@ : What?

W/ : With

W/B : Welcome back

W3 : WWW (Web address)

W8 : Wait

WAH : Working at home

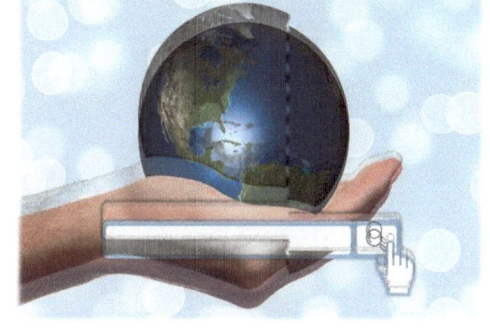

WAJ : What a jerk

WAM : Wait a minute

WAN2 : Want to? (SMS)

WAN2TLK : Want to talk

WAREZ : Pirated (illegally gained) software

WAS : Wait a second

WAS : Wild *a* guess

WAT : What

WAWA : Where are we at?

WAYF : Where are you from?

W/B : Write back

WB : Welcome back

WBS : Write back soon

WBU : What about you?

WE : Welcome

WC : Who cares

WCA : Who cares anyway

W/E : Whatever

W/END : Weekend

WE : Whatever

WEEBO : Describes a person obsessed with Japanese culture

WEP : Weapon (online gaming)

WH5 : Who, what, when, where, why

WHATEVES : Whatever

WIBNI : Wouldn't it be nice if

WDALYIC : Who died and left you in charge

WDYK : What do you know?

WDYT : What do you think?

WGACA : What do you think?

WIIFM : What's in it for me?

WISP : Winning is so pleasurable

WITP : What is the point?

WITW : What in the world

WIU : Wrap it up

WK : Week

WKD : Weekend

WRT : With regard to

WL : What a loser

W/O : Without

WOA : Work of Art

WOKE : Slang for people who are aware of current social issues, and politics

WOMBAT : Waste of money, brains and time

WOW : World of Warcraft (online gaming)

WRK : Work

WRU : Where are you?

WRU@ : Where are you at?

WRUD : What are you doing?

WTB : Want to buy (online gaming)

WTF : What the *freak* ?

WTFE : What the *freak* ever

WTFO : What the *freak* ?, over.

WTG : Way to go

WTGP : Want to go private (talk out of public chat area)

WTH : What the heck?

WTM : Who's the man?

WTS : Want to sell? (online gaming)

WTT : Want to trade? (online gaming)

WU : What's up?

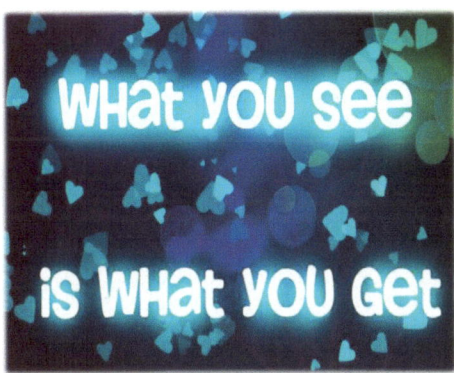

WUCIWUG : What you see is what you get

WUF : Where are you from?

WUP : What's up?

WUT : Meaning "What"

WUW : What u (you) want?

WUU2 : What are you up to?

WUZ : Meaning "Was"

WWJD : What would Jesus do?

WWJD : What Would Judd Do? (Artist Donald Judd, who was known to do things in his own compulsive way)

WWNC : Will wonders never cease

WWYC : Write when you can

WYCM : Will you call me?

WYD : What (are) you doing?

WYGAM : When you get a minute

WYHAM : When you have a minute

WYLEI : When you least expect it

WYSIWYG : What you see is what you get

WYWH : Wish you were here

"X" TEXT MESSAGE & CHAT ABBREVIATIONS

Text Message & Chat Abbreviations

X=1=10 : Meaning "Exciting"

X : Kiss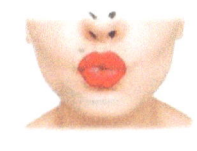

X! : Meaning "A typical woman"

XD : Meaning "Really hard laugh" (where D is a smiley mouth)

XD : Meaning "Devilish smile"

XME : Excuse Me

XOXO : Hugs & Kisses

XLNT : Excellent

XLR8 : Meaning "faster" or "going faster"

XPOST : Meaning "Cross-post" (a link posted to a subreddit that was already shared on a different subreddit)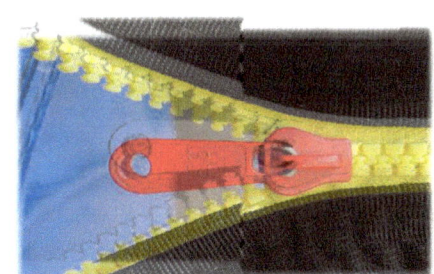

XYL : Ex-young lady (meaning "Wife")

XYZ : Examine your zipper

"Y" TEXT MESSAGE & CHAT ABBREVIATIONS

Y? : Why?

Y : Meaning "Yawn"

Y2K : You're too kind

YA : Your

YAA : Yet another acronym

YABA : Yet another bloody acronym

YARLY : Ya, really?

YAS : Meaning "Praise"

YBIC : Your brother in Christ

YBS : You'll be sorry

YCDBWYCID : You can't do business when your computer is down

YCHT : You can have them

YCLIU : You can look it up

YCMU : You crack me up

YCT : Meaning "Your comment to?"

YD : Yesterday

YEET : Meaning "Excitement, approval or display of energy (i.e. throwing something)"

YF : Wife

YG : Young gentleman

YGG : You go girl

YGTBKM : You've got to be kidding me

YGTR : You got that right

YHBT : You have been trolled

YHBW : You have been warned

YHL : You have lost

YIU : Yes, I understand

YKW : You know what

YKWYCD : You know what you can do

Text Message & Chat Abbreviations

YL : Young lady

YMMV : Your mileage may vary

YNK : You never know

YOLO : You only live once

YR : Your

YR : Yeah right

YRYOCC : You're running your own cuckoo clock

YSIC : Your sister in Christ

YSYD : Yeah sure you do

YT : YouTube

YT : You there?

YTB : You're the best

YTB : Youth talk back

YTTL : You take too long

YTG : You're the greatest

YW : You're welcome

YWHNB : Yes, we have no bananas

YWHOL : Yelling "woohoo" out loud

YWSYLS : You win some, you lose some

YYSSW : Yeah, yeah, sure, sure, whatever

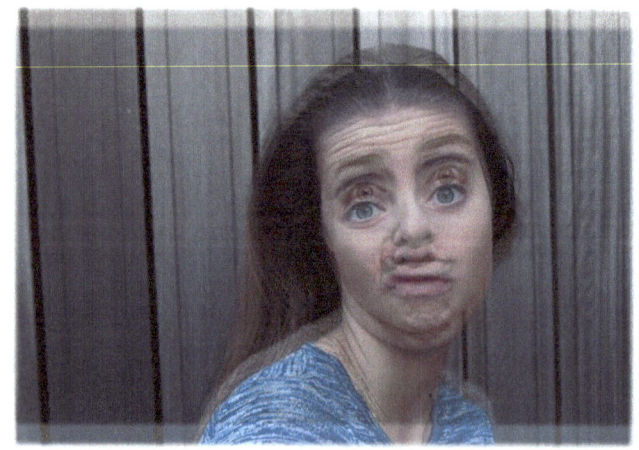

"Z" TEXT MESSAGE & CHAT ABBREVIATIONS

Z : Zero

Z : Z's are calling (meaning "Going to bed/sleep")

Z : Meaning "Said"

Z% : Zoo

ZH : Sleeping Hour

ZOMG : Used in World of Warcraft to mean OMG (Oh My God)

ZOT : Zero tolerance

ZUP : Meaning "What's up?"

ZZZZ : Sleeping (or bored)

NUMBERS & CHARACTERS MESSAGE & CHAT ABBREVIATIONS

Text Message & Chat Abbreviations

? : I have a question

? : I don't understand what you mean

?4U : I have a question for you

;S : Gentle warning, like "Hmm? What did you say?"

AA : Meaning "Read line or read message above"

<3 : Meaning "Sideways heart" (love, friendship)

<33 : Meaning "Heart or love" (more 3s is a bigger heart)

@TEOTD : At the end of the day

:02 : My (or your) two cents worth

1TG, 2TG : Meaning "Number of items needed for win" (online gaming)

1UP : Meaning "Extra life" (online gaming)

121 : One-to-one (private chat initiation)

1337 : Leet (meaning "Elite")

143 : I love you

1432 : I love you, too

14AA41 : One for all, and all for one

182 : I hate you

19 : Zero hand (online gaming)

10M : Ten man (online gaming)

10X : Thanks

10Q : Thank you

1CE : Once

1DR : I wonder

1NAM : One in a million

2 : Meaning "To" in SMS

20 : Meaning "Location"

2B : To be

2EZ : Too easy

2G2BT : Too good to be true

Text Message & Chat Abbreviations

2M2H : Too much to handle
2MI : Too much information
2MOR : Tomorrow
2MORO : Tomorrow
2M2H : Too much to handle
2N8 : Tonight
2NTE : Tonight
4 : Meaning "For" in SMS
404 : I don't know
411 : Meaning "Information"
420 : Let's get high
420 : Meaning "Marijuana"
459 : Means I love you (ILY is 459 using keypad numbers)

4AO : For adults only
4COL : For crying out loud
4EAE : Forever and ever
4EVA : Forever
4NR : Foreigner
4SALE : For sale
^5 : High-five
511 : Too much information (more than 411)
555 : Sobbing, crying (Mandarin Chinese)
55555 : Crying your eyes out (Mandarin Chinese)
55555 : Meaning "Laughing" (In Thai language, "5" is pronounced 'ha')

6Y : Sexy
7K : Sick
81 : Meaning "Hells Angels" (In alphabet, H=8th letter, A=1st letter)
831 : I love you (8 letters, 3 words, 1 meaning)
86 : Over

93

88 : Bye-bye (Mandarin Chinese)

88: Hugs and kisses

9 : Parent is watching

s : Meaning "smile"

w : Meaning "wink"

www.ingramcontent.com/pod-product-compliance
Lightning Source LLC
Chambersburg PA
CBHW051914210526
45473CB00006B/2008